# WHAT IS BODMAS?
## MASTERING THE ORDER OF OPERATIONS
MARIAN PERERA

@MATHISMATHING

@MATHISMATHING

***This book is dedicated to my family, whose unwavering support and inspiration have been truly incredible.***

If you found this book helpful, I would love to hear your thoughts. Please consider leaving a review to share your experience. It would mean a lot and help others discover the book as a useful resource for learning math. Thank you for your support!

@MATHISMATHING

@MATHISMATHING

# Hi, I'm Marian

I'm passionate about math—so much so that I even enrolled in a PhD program in Statistics. As of writing this book, I'm in my third year of the program. I believe math often gets a bad rap for all the wrong reasons. If you think about it, math is like a language—a way to translate the world around us into something we can understand and explore.

Learning math is just like learning any other language. Once you know the words and the rules, you can create and explore something extraordinary and beautiful with it. Math is everywhere—in the patterns of nature, musical rhythms, and buildings' structures. It's a tool that helps us solve problems, make predictions, and see the hidden connections in our world.

I know math can seem intimidating, but it doesn't have to be. Think of it as a puzzle or a game, where each piece you learn brings you closer to seeing the whole picture. It's about curiosity, creativity, and a sense of wonder. When you approach math with an open mind, you'll find that it's not just numbers and formulas—it's a way to unlock the mysteries of the universe.

Let's embark on this journey to learn this incredible language together. We'll take it step by step, helping each other along the way. With patience and practice, we can all become fluent in the language of math and use it to understand the world around us better.

@MATHISMATHING

# TABLE OF CONTENTS

1. What is the Order of Operations ...................... 9
2. Working with Addition and Subtraction ........ 14
3. Working with Division and Multiplication .... 17
4. Working with Brackets ................................. 22
5. Working with Orders ................................... 27
6. Examples ..................................................... 32
7. Your turn ..................................................... 38
8. Answers ...................................................... 51

@MATHISMATHING

@MATHISMATHING

# 1. What is the Order of Operations

***The Order of Operations*** is like a *recipe* for solving any math expression. It makes sure you follow the right steps in the right order. Think of it like baking a cake—you wouldn't put the icing on before the cake is baked, right? BODMAS works in the same way, making sure you handle each part of a math problem in the correct order, so the end result is tasty (or, in this case, correct)!

The secret code to remembering this magic recipe is called **BODMAS**, and here's what it stands for:

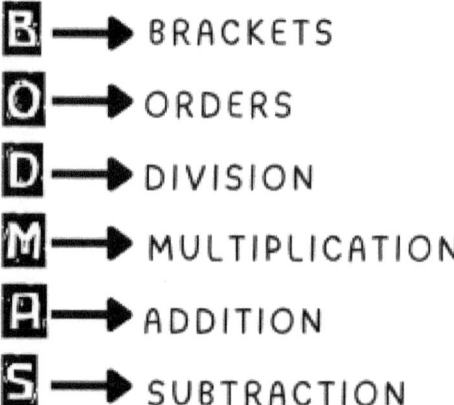

Here's a fun game for you! Try solving this math puzzle in any way you like. Here's your chance to go crazy with Math.

8 + 4 × 2

Did you get 24? Or did you get 16?

@MATHISMATHING

If you start from the left and go straight across, you might think it's:

$(8 + 4) \times 2 = 12 \times 2 = 24$

But wait! BODMAS says, "Hold on! You need to multiply first!" So, it should be:

$8 + (4 \times 2) = 8 + 8 = 16$

See the difference? This is why **the Order of Operations** is important. It is like a referee in a game, making sure all the math moves are fair and square. Without it, people could solve problems any way they wanted, and no one would agree on the answers. That would be chaos—like a soccer match where everyone runs in different directions!

So, when we have a bunch of math operations to solve, here's what we do: start with anything inside brackets, then handle orders. Next, work through division and multiplication from left to right, and finally, tackle addition and subtraction. By following this order, we can ensure our math 'recipe' turns out perfectly every time, without any surprises!

@MATHISMATHING

## Let's try solving,

$$6 + 2 \times (3^2 - 1)$$

**B:** The brackets
$(3^2 - 1) = (9 - 1) = 8.$
    Now
    $6 + 2 \times (3^2 - 1) = 6 + 2 \times 8.$

**O:** We already handled in the orders inside the brackets, so we can move on.

**D:** There's no division here, so we go to the next step.

**M:** The multiplication
$2 \times 8 = 16.$
    Now
$6 + 2 \times 8 = 6 + 16.$

**A:** The addition
$6 + 16 = 22.$

**S:** The subtraction
    There's no subtraction.

@MATHISMATHING

## So, the final answer is:

$$6 + 2 \times (3^2 - 1) = 22$$

Get ready for a math adventure as we explore each component of BODMAS in detail in the upcoming chapters! Instead of tackling them in the order they appear, we'll follow the path of simplicity, starting with the easiest concepts. So, buckle up, grab your math cap, and let's leap into this exciting journey together! Ready, set, let's go!

@MATHISMATHING

# 2. Working with Addition and Subtraction

Just like multiplication and division, **addition and subtraction are a perfect pair**, working together like math's dynamic duo! They're opposites, but that's what makes them such a great team—**addition puts things together**, while **subtraction takes them apart**. Just like multiplication and division, addition and subtraction are also **inverse operations**. This means they balance each other out, making them the perfect pair in the math universe. So, whether you're stacking up numbers or taking them down, these two operations are always ready to team up and tackle any challenge!

## Addition

Addition is like throwing a party where you invite numbers to come together and make a bigger number! When you add, you're simply combining quantities to find out how much you have in total.

For example, if you have $3$ apples and you add 2 more, you now have $3 + 2 = 5$ apples!

In math, the plus sign (+) is the magic symbol that tells you to join numbers. You can think of it as the friendly handshake that brings numbers together to create a new total.

@MATHISMATHING

# Subtraction

**Subtraction** is like a magician making numbers disappear! When you subtract, you're taking away a quantity from another to see what's left.

For example, if you start with 5 candies and you eat 2, you have 5−2=3 candies left.

In math, the minus sign (-) is the symbol that indicates you're removing something. It's like waving a magic wand to make some numbers vanish!

# Order of Operations

Even though according to BODMAS, both addition and subtraction have the same level of priority. This means you simply work from **left to right**.

For example:
8 − 3 + 2
Even though A comes first before D in BODMAS, the subtraction goes first because when going from left to right the first one we have is subtraction.

### Step 1: Starting with subtraction
8 − 3 = 5
      Now the expression would be
5 + 2

### Step 2: It's time for the addition
5 + 2 = 7

@MATHISMATHING

7 is our answer!

# 3. Working with Division and Multiplication

These two operations are like the inseparable friends of the math world, always complementing each other. In fact, any division can be rewritten as a multiplication, and vice versa. It's like they speak the same language, just in different ways.

## Multiplication

Multiplication is all about **repeated addition**. When you multiply, you're essentially saying, "I want to add this number to itself a certain number of times."

For example:
- 4×3 means you're adding 4 three times: 4+4+4=12.

You can think of multiplication as a shortcut to add numbers quickly. If you had 4 pizzas with 3 slices each, multiplication helps you find out that you have a total of 4×3=12 slices without counting them one by one!

## Division

Division is all about sharing or breaking down a number into equal parts. When you divide, you're asking, "How many times does this number fit into that one?"

@MATHISMATHING

For example:

- 12÷3 means you're figuring out how many times 3 fits into 12, which is 4.

Imagine you have 12 cookies, and you want to share them equally among 3 friends. Each friend would get 12÷3=4 cookies. Division helps you distribute things fairly, making sure everyone gets their share!

# The Relationship Between Multiplication and Division

Multiplication and division are closely connected, like two sides of the same coin. They are **inverse operations**, meaning that they can undo each other. For example, if you know that 4×3=12, then you can also say 12÷4=3 and 12÷3=4.

# Order of Operations

It can get a little tricky when multiplication and division appear together in an expression, especially when there are no brackets to separate them. Even though **"D"** comes before **"M"** in **BODMAS**, there isn't a strict rule that division must always go first. Since multiplication and division are **inverse operations**, the key is to handle them in the order they appear from **left to right**.

So, when you encounter both operations in an expression, work from left to right. If you come across multiplication before division, go ahead and multiply first. If division shows up first, then divide. It doesn't matter that **"M"** comes after **"D"** in **BODMAS**—just follow the sequence from left to right, and you'll always get it right!

For example:
$8 + 4 \times 2 \div 2$
we need to take care of multiplication and division first before dealing with addition.

**Step 1:** Going from left to right we got multiplication first.
$4 \times 2 = 8$
    Then the expression becomes
    $8 + 8 \div 2$

**Step 2:** Now we deal with the division.
$8 \div 2 = 4$
    So, the expression becomes:
$8 + 4$

@MATHISMATHING

**Step 3:** Finally, we got the addition.

8 + 4 = 12

So, the answer is **12!**

# 4. Working with Brackets

***Brackets*** are like the first step in solving any math puzzle—they're the wrapping paper on a present that you have to open before seeing what's inside! In math, brackets tell you which parts of an expression to solve first; they separate parts of an equation and show you exactly where to start. Without brackets, math problems could have different answers depending on the order in which you solve them. Brackets ensure that everyone does the math in the same sequence and arrives at the same correct answer.

There are different types of brackets, each serving the same purpose: to group things together and make math easier to handle. There are three main types of brackets you'll come across in math,

- **Parentheses ( )**: The most common type, like the basic wrapping paper. They look like curved lines and are used to group things together, such as (3+2).

- **Square Brackets [ ]**: A bit sturdier, like a box inside the wrapping paper. Usually they are used when there are already parentheses in the expression, such as [3×(2+1)].

- **Curly Braces { }**: The fancy kind of wrapping used for larger expressions or when there are both parentheses and square brackets already, like {5+[(2+1)÷2]}.

You solve from the **inside out**: start with the innermost brackets and work your way out.

But here's the fun twist: you don't always have to use a mix of brackets! Some math wizards prefer to stick with

@MATHISMATHING

just parentheses, while others might use a combination of two or even all three types, mixing them up in a delightful math salad!

The key to navigating this bracket party is to keep track of the pairs—every opening bracket has a closing bracket! A super simple way to remember this is the **"Last In, First Out"** rule, like a stack of plates. The last plate you stack on top is the first one you take off! So, when you're solving math problems, just remember: the last bracket you opened is the first one you'll close.

See if you can identify the pairs. Remember the rule **"Last In, First Out"**!

@MATHISMATHING

# Let's practice, shall we?

$2 \times (3 + 4)$

## Step 1: Start with the brackets.

$3 + 4 = 7$

     Now the expression is:

$2 \times 7.$

## Step 2: Then the multiplication.

$2 \times 7 = 14$

And that is how you do it! **14** is your answer.

@MATHISMATHING

## Let's do another one.

$((8 \times 2) - 10) \div 2$

**Step 1:** Start with the innermost brackets.

$(8 \times 2) = 16$

   Now we have

$(16 - 10) \div 2.$

**Step 2**: Now, we can do the brackets.

$(16 - 10) = 6$

   Then the expression becomes:

$6 \div 2 = 3.$

**3** is your answer!

@MATHISMATHING

# 5. Working with Orders

Welcome to the magical world of orders, where numbers can grow, shrink, and transform in amazing ways! Think of these concepts as the math equivalent of casting spells—raising numbers to great heights, shrinking them down, or finding their hidden secrets. In this chapter, we'll explore two main types of orders: the ones that expand numbers like a balloon and the ones that shrink them like a deflating airbag. These magical powers are known as **powers** (or exponents) and **roots**.

## Powers

Powers (or exponents) are like giving numbers a boost of super strength! When you see a small number floating above a big number, like $3^2$, it means you're multiplying that number by itself a certain number of times.

For example:

- $3^2$ (read as "three squared") means 3×3.

- $5^3$ (read as "five cubed") means 5×5×5.

In general, when you see $a^n$, it means you multiply *a* by itself *n* times. Think of the exponent as the number of times you wave your magic wand to make the base number grow stronger!

@MATHISMATHING

# Roots

Roots are like reverse magic—shrinking numbers back down to their original form. The most common type is the square root, written as $\sqrt[2]{25}$, which asks, "What number multiplied by itself equals 25?" In this case, the answer is 5 because 5×5=25. The square root is so popular that we often ditch that little 2 sitting up high, like a shy friend who prefers to hang back! So, whether you see $\sqrt[2]{25}$ or just $\sqrt{25}$, they're actually the same party guest—they both invite you to find the number that, when multiplied by itself, equals 25.

That little number you see up there is like a math clue—it tells you how many times a number should be multiplied by itself to reach the big number. For example, when you see $\sqrt[3]{27}$, it's asking, "What's the magic number you need to multiply three times to get 27?" It's like a math scavenger hunt where the tiny number on top tells you how many steps it takes to find the treasure. In this case, the answer is 3, because 3×3×3=27.

@MATHISMATHING

# Here's an example

$2^3 + \sqrt{16}$

**Step 1:** Start with the powers and roots.

$2^3 = 2 \times 2 \times 2 = 8$
and
$\sqrt{16} = \sqrt{4 \times 4} = 4.$
So the expression is:
8 + 4.

**Step 2:** Now we can add.

8 + 4 = 12.

So your answer is **12!**

@MATHISMATHING

# Let's do another one.

$(3^2 + 4) \times \sqrt{9}$

**Step 1:** Work inside the brackets first.
$(3^2 + 4) = (3 \times 3 + 4) = 9 + 4 = 13$
    Now we got
$13 \times \sqrt{9}$.

**Step 2:** Solve the square root
$\sqrt{9} = 3$
        Then the expression becomes
                $13 \times 3$.

**Step 3:** Finally, we can do the multiplication.

            $13 \times 3 = 39$.

So the answer is **39**!

@MATHISMATHING

# 6.Examples

Now that we've explored each component of BODMAS, it's time to put it all into action and make sense of how it works together. Let's see how the rules come to life when we tackle real expressions!

Following this chapter, you'll find some worked examples to guide you through the process, followed by practice problems for you to solve on your own.

# $20 \div 5 + 6 \times 2$

**B:** The brackets

    No brackets, so we move to the next step.

**O:** The orders

    There's no order, let's jump to the next step.

**D:** The division

$20 \div 5 = 4$.

    Then

$20 \div 5 + 6 \times 2 = 4 + 6 \times 2$.

**M:** Multiplication next

$6 \times 2 = 12$

    Then

$4 + 6 \times 2 = 4 + 12$

**A:** Addition next

$4 + 12 = 16$

**S:** No subtraction.

**So, the final answer is:**

$$20 \div 5 + 6 \times 2 = 16$$

@MATHISMATHING

# $(10 \div 2) \times (3 + 1)$

**B:** Let's start with the brackets.
$10 \div 2 = 5$
$3 + 1 = 4$
    So
    $(10 \div 2) \times (3 + 1) = 5 \times 4.$

**O:** No orders.

**D:** Already handled division.

**M:** Let's do the multiplication
$5 \times 4 = 20$

**A:** No addition

**S:** No subtraction

**So, the final answer is:**

$$(10 \div 2) \times (3 + 1) = 20$$

@MATHISMATHING

$$3^3 \div (6 + 3)$$

**B:** Let's start with the brackets.
$6 + 3 = 9$
    So
    $3^3 \div (6 + 3) = 3^3 \div 9.$

**O:** Now the orders.
$3^3 = 3 \times 3 \times 3 = 27$
    Now
$3^3 \div 9 = 27 \div 9$

**D:** Next the division.
$27 \div 9 = 3$

**M:** No multiplication

**A:** Already took care of addition

**S:** No subtraction

**So, the final answer is:**
$$3^3 \div (6 + 3) = 3$$

@MATHISMATHING

$$5 \times 2 \div (2^2 - 1)$$

**B:** Let's start with the brackets.
$2^2 - 1 = 4 - 1 = 3$
　　So
　　$5 \times 2 \div (2^2 - 1) = 5 \times 2 \div 3.$

**O:** No orders left

**M:** Now the multiplication because multiplication comes first before division when going from left to right.
$5 \times 2 = 10$
　　So
$5 \times 2 \div 3 = 10 \div 3$

**D:** Let's do the division.
$10 \div 3 = 3.3333$

**A:** No addition

**S:** No subtraction left

**So, the final answer is:**

$$5 \times 2 \div (2^2 - 1) = 3.3333$$

@MATHISMATHING

# 7. Your turn

1. $8 + 2 \times (6 - 3)$

2. $(5 + 3) \times 4 - 6$

3. $18 \div (3 + 3) + 2 \times 5$

4. $7 \times (5 + 2) - 9 \div 3$

5. $(3^2 + 4) \div 2 - 1$

6. $2 \times (3 + 7) - 4^2$

7. $9 \times (8 - 4) \div 2 + 15$

@MATHISMATHING

8. $(6 + 2)^2 \div 4 \times 3$

9. $14 - 3 \times (4 + 2)$

10. $25 \div (3 + 2) \times 4$

11. $2 + (8 \div 4) \times 5$

12. $10 - 3 + (8 \div 2)$

13. $(5 + 9) \times 3 \div (6 - 2)$

14. $12 \times 2 \div (3 + 1)$

15. $(18 + 6) \div 3 + 2^3$

@MATHISMATHING

16.     $4^2 - (3 + 5) \times 2$

17.     $(5 + 2)^3 \div 7$

18.     $3 \times 4^2 \div 8 + 7$

19.     $9 \div 3 \times (6 + 1)$

20.     $14 + 8 \div 4 \times 2$

21.     $30 - (5 \times 2) + 6^2$

22.     $(8 \times 5) \div 10 + 4^2$

23.     $7 + 2 \times (10 \div 5)$

@MATHISMATHING

**24.** $(6 \div 2 + 3) \times 4$

**25.** $15 + 3 \times 2^2$

**26.** $(10 - 4)^3 \div 9$

**27.** $5^2 - (4 + 6) \div 2$

**28.** $7 \times (3 + 2) \div 5$

**29.** $11 + 3^2 - 8$

**30.** $9 \times (5 + 4) \div 3$

**31.** $(16 \div 4)^2 + 3$

@MATHISMATHING

32. $18 + 4 \div 2 \times 3$

33. $21 - (2 + 3) \times 5$

34. $(3 \times 6) + 5 \div 2$

35. $7^2 \div (4 + 3)$

36. $(8 + 5) \times 2 - 3^2$

37. $6 \div 3 \times (2 + 4)$

38. $20 \div (3 + 2) + 5^2$

39. $(15 - 5) \times 3 + 2^2$

@MATHISMATHING

40. $2^3 \times 4 - (6 + 2)$

41. $7 + 2 \times (6 - 4)^2$

42. $30 \div (6 + 4) \times 3$

43. $(12 - 4)^2 \div 8$

44. $25 + (6 \div 2) \times 3$

45. $3 \times (5 + 2) \div 7$

46. $(8 + 4) \times 3 \div 6$

47. $10 - (3^2 + 4)$

@MATHISMATHING

**48.** $9 \times 2^2 - (5 + 3)$

**49.** $(15 + 5) \div 4 \times 2$

**50.** $6^2 - (3 \times 4) + 2$

**51.** $3 \times (7 + 2) - 4^2$

**52.** $8 \div (4 + 2) \times 5$

**53.** $(10 \times 3) + 2 \div 5$

**54.** $18 \div (2 + 4) \times 3$

**55.** $(7 - 3) \times 5 + 2^3$

@MATHISMATHING

**56.** $12 + (4 \times 2)^2 \div 8$

**57.** $5 \times 6 - (2 + 3)^2$

**58.** $(9 \div 3 + 4) \times 2$

**59.** $4^3 - (8 \div 4) \times 5$

**60.** $25 \div (3 + 2) + 3^2$

**61.** $(14 + 6) \div 2 \times 4$

**62.** $8 \times 3 \div (7 - 4)$

**63.** $(5 + 2)^2 \div 9$

@MATHISMATHING

64. $3^3 - (7 + 5) \times 2$

65. $12 \div (4 + 2) \times 5$

66. $(8 \times 2) + 3 \div 7$

67. $10 + (5 \div 1) \times 2^2$

68. $(7 \times 4) - 5^2$

69. $3 + (4 \times 5) \div 2$

70. $(9 - 3) \times 4^2$

71. $16 + (3 \div 1) \times 2^3$

@MATHISMATHING

72. $(6 \times 4) + 3^2 - 8$

73. $14 \div (3 + 4) \times 5$

74. $(2^3 + 7) \div 3$

75. $10 \times (5 - 3)^2$

76. $(8 + 6) \div 7 + 3^2$

77. $3 \times (9 + 1) \div 5$

78. $(12 - 5) \times 4 + 2^2$

79. $20 \div (2 + 3)^2$

@MATHISMATHING

80. $(6 + 4)^2 \div 2$

81. $11 - (3 \times 2)^2 \div 6$

82. $(8 \div 4) \times 3 + 7$

83. $9 + (4 \times 2) - 3^2$

84. $(5 \times 3) + 2 \div 4$

85. $14 \div (7 - 5) + 2^2$

86. $(6 \times 2)^2 \div 8$

87. $3 \times (5 + 4)^2 \div 6$

@MATHISMATHING

**88.**     $(10 ÷ 2) + 3^3$

**89.**     $12 - (4 × 3) + 2^2$

**90.**     $(7 × 2) ÷ 4 + 3$

**91.**     $16 ÷ (8 + 4) × 9$

**92.**     $(6 + 3) × 2 - 7^2$

**93.**     $18 ÷ (9 - 3) + 5^2$

**94.**     $(8 - 4)^3 ÷ 2$

**95.**     $25 × 3^2 - (7 + 6)$

@MATHISMATHING

**96.**     $(5 \div 1) + 2^3 \times 4$

**97.**     $10 - (3 \times 2) + 4^2$

**98.**     $(9 \times 3) - 5 \div 2$

**99.**     $7 + (8 \div 4)^2$

**100.**    $(6 \times 3) + 2^3 \div 4$

@MATHISMATHING

## 8. Answers

**1.** $8 + 2 \times (6 - 3)$
$= 8 + 2 \times 3$
$= 8 + 6$
$= 14$

**2.** $(5 + 3) \times 4 - 6$
$= 8 \times 4 - 6$
$= 32 - 6$
$= 26$

**3.** $18 \div (3 + 3) + 2 \times 5$
$= 18 \div (3 + 3) + 2 \times 5$
$= 18 \div 6 + 2 \times 5$
$= 3 + 2 \times 5$
$= 3 + 10$
$= 13$

@MATHISMATHING

**4.** $7 \times (5 + 2) - 9 \div 3$
$= 7 \times 7 - 9 \div 3$
$= 49 - 9 \div 3$
$= 49 - 3$
$= 46$

**5.** $(3^2 + 4) \div 2 - 1$
$= (3^2 + 4) \div 2 - 1$
$= (9 + 4) \div 2 - 1$
$= 13 \div 2 - 1$
$= 6.5 - 1$
$= 5.5$

**6.** $2 \times (3 + 7) - 4^2$
$= 2 \times 10 - 4^2$
$= 2 \times 10 - 16$
$= 20 - 16$
$= 4$

@MATHISMATHING

7. $9 \times (8 - 4) \div 2 + 15$
   $= 9 \times 4 \div 2 + 15$
   $= 36 \div 2 + 15$
   $= 18 + 15$
   $= 33$

8. $(6 + 2)^2 \div 4 \times 3$
   $= (6 + 2)^2 \div 4 \times 3$
   $= 8^2 \div 4 \times 3$
   $= 64 \div 4 \times 3$
   $= 16 \times 3$
   $= 48$

9. $14 - 3 \times (4 + 2)$
   $= 14 - 3 \times 6$
   $= 14 - 18$
   $= -4$

10. $25 \div (3 + 2) \times 4$
    $= 25 \div 5 \times 4$
    $= 5 \times 4$
    $= 20$

@MATHISMATHING

**11.** $2 + (8 \div 4) \times 5$
$= 2 + (8 \div 4) \times 5$
$= 2 + 2 \times 5$
$= 2 + 10$
$= 12$

**12.** $10 - 3 + (8 \div 2)$
$= 10 - 3 + 4$
$= 7 + 4$
$= 11$

**13.** $(5 + 9) \times 3 \div (6 - 2)$
$= 14 \times 3 \div 4$
$= 42 \div 4$
$= 10.5$

**14.** $12 \times 2 \div (3 + 1)$
$= 12 \times 2 \div (3 + 1)$
$= 12 \times 2 \div 4$
$= 24 \div 4$
$= 6$

@MATHISMATHING

**15.** $(18 + 6) \div 3 + 2^3$
$\phantom{15.}= (18 + 6) \div 3 + 2^3$
$\phantom{15.}= 24 \div 3 + 2^3$
$\phantom{15.}= 24 \div 3 + 8$
$\phantom{15.}= 8 + 8$
$\phantom{15.}= 16$

**16.** $4^2 - (3 + 5) \times 2$
$\phantom{16.}= 4^2 - 8 \times 2$
$\phantom{16.}= 16 - 8 \times 2$
$\phantom{16.}= 16 - 16$
$\phantom{16.}= 0$

**17.** $(5 + 2)^3 \div 7$
$\phantom{17.}= 7^3 \div 7$
$\phantom{17.}= 343 \div 7$
$\phantom{17.}= 49$

**18.** $3 \times 4^2 \div 8 + 7$
$\phantom{18.}= 3 \times 16 \div 8 + 7$
$\phantom{18.}= 6 + 7$
$\phantom{18.}= 13$

@MATHISMATHING

**19.** $9 \div 3 \times (6 + 1)$
$= 9 \div 3 \times 7$
$= 3 \times 7$
$= 21$

**20.** $14 + 8 \div 4 \times 2$
$= 14 + 2 \times 2$
$= 14 + 4$
$= 18$

**21.** $30 - (5 \times 2) + 6^2$
$= 30 - 10 + 6^2$
$= 30 - 10 + 36$
$= 20 + 36$
$= 56$

**22.** $(8 \times 5) \div 10 + 4^2$
$= 40 \div 10 + 4^2$
$= 40 \div 10 + 16$
$= 4 + 16$
$= 20$

@MATHISMATHING

**23.** $7 + 2 \times (10 \div 5)$
$= 7 + 2 \times 2$
$= 7 + 4$
$= 11$

**24.** $(6 \div 2 + 3) \times 4$
$= (3 + 3) \times 4$
$= 6 \times 4$
$= 24$

**25.** $15 + 3 \times 2^2$
$= 15 + 3 \times 4$
$= 15 + 12$
$= 27$

**26.** $(10 - 4)^3 \div 9$
$= (10 - 4)^3 \div 9$
$= 6^3 \div 9$
$= 24$

@MATHISMATHING

**27.** $5^2 - (4 + 6) \div 2$

$= 5^2 - 10 \div 2$

$= 25 - 10 \div 2$

$= 25 - 5$

$= 20$

**28.** $7 \times (3 + 2) \div 5$

$= 7 \times 5 \div 5$

$= 35 \div 5$

$= 7$

**29.** $11 + 3^2 - 8$

$= 11 + 9 - 8$

$= 20 - 8$

$= 12$

**30.** $9 \times (5 + 4) \div 3$

$= 9 \times (5 + 4) \div 3$

$= 9 \times 9 \div 3$

$= 81 \div 3$

$= 27$

@MATHISMATHING

**31.** $(16 \div 4)^2 + 3$
$\phantom{31.}= 4^2 + 3$
$\phantom{31.}= 16 + 3$
$\phantom{31.}= 19$

**32.** $18 + 4 \div 2 \times 3$
$\phantom{32.}= 18 + 2 \times 3$
$\phantom{32.}= 18 + 6$
$\phantom{32.}= 24$

**33.** $21 - (2 + 3) \times 5$
$\phantom{33.}= 21 - 5 \times 5$
$\phantom{33.}= 21 - 25$
$\phantom{33.}= -4$

**34.** $(3 \times 6) + 5 \div 2$
$\phantom{34.}= 18 + 5 \div 2$
$\phantom{34.}= 18 + 2.5$
$\phantom{34.}= 20.5$

**35.** $7^2 \div (4 + 3)$
$= 7^2 \div 7$
$= 49 \div 7$
$= 7$

**36.** $(8 + 5) \times 2 - 3^2$
$= 13 \times 2 - 3^2$
$= 13 \times 2 - 9$
$= 26 - 9$
$= 17$

**37.** $6 \div 3 \times (2 + 4)$
$= 6 \div 3 \times 6$
$= 2 \times 6$
$= 12$

**38.** $20 \div (3 + 2) + 5^2$
$= 20 \div 5 + 5^2$
$= 20 \div 5 + 25$
$= 4 + 25$
$= 29$

@MATHISMATHING

**39.** $(15 - 5) \times 3 + 2^2$
$= 10 \times 3 + 2^2$
$= 10 \times 3 + 4$
$= 30 + 4$
$= 34$

**40.** $2^3 \times 4 - (6 + 2)$
$= 2^3 \times 4 - 8$
$= 8 \times 4 - 8$
$= 32 - 8$
$= 24$

**41.** $7 + 2 \times (6 - 4)^2$
$= 7 + 2 \times 2^2$
$= 7 + 2 \times 4$
$= 7 + 8$
$= 15$

**42.** $30 \div (6 + 4) \times 3$
$= 30 \div 10 \times 3$
$= 3 \times 3$
$= 9$

@MATHISMATHING

**43.** $(12 - 4)^2 \div 8$
$= 8^2 \div 8$
$= 64 \div 8$
$= 8$

**44.** $25 + (6 \div 2) \times 3$
$= 25 + (6 \div 2) \times 3$
$= 25 + 3 \times 3$
$= 25 + 9$
$= 34$

**45.** $3 \times (5 + 2) \div 7$
$= 3 \times 7 \div 7$
$= 3 \times 1$
$= 3$

**46.** $(8 + 4) \times 3 \div 6$
$= (8 + 4) \times 3 \div 6$
$= 12 \times 3 \div 6$
$= 36 \div 6$
$= 6$

**47.** $10 - (3^2 + 4)$
$= 10 - (9 + 4)$
$= 10 - 13$
$= -3$

**48.** $9 \times 2^2 - (5 + 3)$
$= 9 \times 2^2 - 8$
$= 9 \times 4 - 8$
$= 36 - 8$
$= 28$

**49.** $(15 + 5) \div 4 \times 2$
$= 20 \div 4 \times 2$
$= 5 \times 2$
$= 10$

**50.** $6^2 - (3 \times 4) + 2$
$= 6^2 - (3 \times 4) + 2$
$= 6^2 - 12 + 2$
$= 36 - 12 + 2$
$= 24 + 2$
$= 26$

@MATHISMATHING

**51.** $3 \times (7 + 2) - 4^2$
$= 3 \times 9 - 4^2$
$= 3 \times 9 - 16$
$= 27 - 16$
$= 11$

**52.** $8 \div (4 + 2) \times 5$
$= 8 \div 6 \times 5$
$= 1.3333 \times 5$
$= 6.6667$

**53.** $(10 \times 3) + 2 \div 5$
$= 30 + 2 \div 5$
$= 30 + 0.4$
$= 30.4$

**54.** $18 \div (2 + 4) \times 3$
$= 18 \div 6 \times 3$
$= 3 \times 3$
$= 9$

@MATHISMATHING

**55.** $(7 - 3) \times 5 + 2^3$
$= 4 \times 5 + 2^3$
$= 4 \times 5 + 8$
$= 28$

**56.** $12 + (4 \times 2)^2 \div 8$
$= 12 + 8^2 \div 8$
$= 12 + 64 \div 8$
$= 12 + 8$
$= 20$

**57.** $5 \times 6 - (2 + 3)^2$
$= 5 \times 6 - 5^2$
$= 30 - 5^2$
$= 30 - 25$
$= 5$

**58.** $(9 \div 3 + 4) \times 2$
$= (3 + 4) \times 2$
$= 7 \times 2$
$= 14$

@MATHISMATHING

**59.** $4^3 - (8 \div 4) \times 5$
$\phantom{59.} = 4^3 - 2 \times 5$
$\phantom{59.} = 64 - 2 \times 5$
$\phantom{59.} = 64 - 10$
$\phantom{59.} = 54$

**60.** $25 \div (3 + 2) + 3^2$
$\phantom{60.} = 25 \div 5 + 3^2$
$\phantom{60.} = 25 \div 5 + 9$
$\phantom{60.} = 5 + 9$
$\phantom{60.} = 14$

**61.** $(14 + 6) \div 2 \times 4$
$\phantom{61.} = 20 \div 2 \times 4$
$\phantom{61.} = 10 \times 4$
$\phantom{61.} = 40$

**62.** $8 \times 3 \div (7 - 4)$
$\phantom{62.} = 8 \times 3 \div (7 - 4)$
$\phantom{62.} = 8 \times 3 \div 3$
$\phantom{62.} = 24 \div 3$
$\phantom{62.} = 8$

@MATHISMATHING

**63.** $(5 + 2)^2 \div 9$
$= 7^2 \div 9$
$= 49 \div 9$
$= 5.44$

**64.** $3^3 - (7 + 5) \times 2$
$= 3^3 - 12 \times 2$
$= 27 - 12 \times 2$
$= 3$

**65.** $12 \div (4 + 2) \times 5$
$= 12 \div 6 \times 5$
$= 2 \times 5$
$= 10$

**66.** $(8 \times 2) + 3 \div 7$
$= 16 + 3 \div 7$
$= 16 + 0.4286$
$= 16.4286$

**67.** $10 + (5 \div 1) \times 2^2$
$= 10 + 5 \times 2^2$
$= 10 + 5 \times 4$
$= 10 + 20$
$= 30$

**68.** $(7 \times 4) - 5^2$
$= 28 - 5^2$
$= 28 - 25$
$= 3$

**69.** $3 + (4 \times 5) \div 2$
$= 3 + 20 \div 2$
$= 3 + 10$
$= 13$

**70.** $(9 - 3) \times 4^2$
$= 6 \times 4^2$
$= 6 \times 16$
$= 96$

@MATHISMATHING

**71.** $16 + (3 \div 1) \times 2^3$
$= 16 + 3 \times 2^3$
$= 16 + 3 \times 8$
$= 16 + 24$
$= 40$

**72.** $(6 \times 4) + 3^2 - 8$
$= 24 + 3^2 - 8$
$= 24 + 9 - 8$
$= 33 - 8$
$= 25$

**73.** $14 \div (3 + 4) \times 5$
$= 14 \div 7 \times 5$
$= 2 \times 5$
$= 10$

@MATHISMATHING

**74.** $(2^3 + 7) \div 3$
$= (8 + 7) \div 3$
$= 15 \div 3$
$= 5$

**75.** $10 \times (5 - 3)^2$
$= 10 \times 2^2$
$= 10 \times 4$
$= 40$

**76.** $(8 + 6) \div 7 + 3^2$
$= 14 \div 7 + 3^2$
$= 14 \div 7 + 9$
$= 2 + 9$
$= 11$

**77.** $3 \times (9 + 1) \div 5$
$= 3 \times 10 \div 5$
$= 30 \div 5$
$= 6$

@MATHISMATHING

**78.** $(12 - 5) \times 4 + 2^2$
$= 7 \times 4 + 2^2$
$= 7 \times 4 + 4$
$= 28 + 4$
$= 32$

**79.** $20 \div (2 + 3)^2$
$= 20 \div 5^2$
$= 20 \div 25$
$= 0.8$

**80.** $(6 + 4)^2 \div 2$
$= 10^2 \div 2$
$= 100 \div 2$
$= 50$

**81.** $11 - (3 \times 2)^2 \div 6$
$= 11 - 6^2 \div 6$
$= 11 - 36 \div 6$
$= 11 - 6$
$= 5$

@MATHISMATHING

**82.** $(8 ÷ 4) × 3 + 7$
$= 2 × 3 + 7$
$= 6 + 7$
$= 13$

**83.** $9 + (4 × 2) - 3^2$
$= 9 + 8 - 3^2$
$= 9 + 8 - 9$
$= 17 - 9$
$= 8$

**84.** $(5 × 3) + 2 ÷ 4$
$= 15 + 2 ÷ 4$
$= 15 + 0.5$
$= 15.5$

**85.** $14 ÷ (7 - 5) + 2^2$
$= 14 ÷ 2 + 2^2$
$= 7 + 2^2$
$= 7 + 4$
$= 11$

@MATHISMATHING

**86.** $(6 \times 2)^2 \div 8$
$= 12^2 \div 8$
$= 144 \div 8$
$= 18$

**87.** $3 \times (5 + 4)^2 \div 6$
$= 3 \times 9^2 \div 6$
$= 3 \times 81 \div 6$
$= 243 \div 6$
$= 40.5$

**88.** $(10 \div 2) + 3^3$
$= (10 \div 2) + 3^3$
$= 5 + 27$
$= 32$

**89.** $12 - (4 \times 3) + 2^2$
$= 12 - 12 + 2^2$
$= 12 - 12 + 4$
$= 0 + 4$
$= 4$

@MATHISMATHING

**90.** $(7 \times 2) \div 4 + 3$
$= 14 \div 4 + 3$
$= 3.5 + 3$
$= 6.5$

**91.** $16 \div (8 + 4) \times 9$
$= 16 \div 12 \times 9$
$= 1.333 \times 9$
$= 12$

**92.** $(6 + 3) \times 2 - 7^2$
$= 9 \times 2 - 7^2$
$= 9 \times 2 - 49$
$= 18 - 49$
$= -31$

**93.** $18 \div (9 - 3) + 5^2$
$= 18 \div 6 + 5^2$
$= 3 + 5^2$
$= 3 + 25$
$= 28$

@MATHISMATHING

**94.** $(8 - 4)^3 \div 2$
$= 4^3 \div 2$
$= 64 \div 2$
$= 32$

**95.** $25 \times 3^2 - (7 + 6)$
$= 25 \times 3^2 - 13$
$= 25 \times 9 - 13$
$= 225 - 13$
$= 212$

**96.** $(5 \div 1) + 2^3 \times 4$
$= 5 + 2^3 \times 4$
$= 5 + 8 \times 4$
$= 5 + 32$
$= 37$

**97.** $10 - (3 \times 2) + 4^2$
$= 10 - 6 + 4^2$
$= 10 - 6 + 16$
$= 4 + 16$
$= 20$

@MATHISMATHING

**98.** $(9 \times 3) - 5 \div 2$
$= 27 - 5 \div 2$
$= 27 - 2.5$
$= 24.5$

**99.** $7 + (8 \div 4)^2$
$= 7 + 2^2$
$= 7 + 4$
$= 11$

**100.** $(6 \times 3) + 2^3 \div 4$
$= 18 + 2^3 \div 4$
$= 18 + 8 \div 4$
$= 18 + 2$
$= 20$

@MATHISMATHING

# Congratulations, Math Master!

If you've made it this far, you're nothing short of amazing! Seriously, **you're insanely fabulous**. Thank you so much for sticking with me throughout this journey of mastering BODMAS and the magical world of math. You've got the spirit and determination to keep pushing forward, and I promise that if you maintain this momentum, **you'll be a math pro in no time.**

I hope you enjoyed the book and found it helpful. If you found this book helpful, I'd be incredibly grateful if you could take a moment to **leave a review.** Your feedback means the world to me and helps others discover the book and join the math adventure. Don't forget to **check out my other books**—they're packed with more tricks and tips to sharpen your math skills even further.

**Kudos to you for your hard work and dedication!** Keep going, keep growing, and always keep that can-do attitude.

Best wishes and happy problem-solving,

**Marian Perera**

@MATHISMATHING

@MATHISMATHING

www.ingramcontent.com/pod-product-compliance
Lightning Source LLC
Chambersburg PA
CBHW070409230526
45471CB00006B/2716